Maryna Psol, Vincenzo Romaniello

Effect of E65 and E70 splice isoforms on electrophysiological properties of Kv10.1

GRIN Verlag

Bibliografische Information der Deutschen Nationalbibliothek:

Die Deutsche Bibliothek verzeichnet diese Publikation in der Deutschen National-
bibliografie; detaillierte bibliografische Daten sind im Internet über http://dnb.d-
nb.de/ abrufbar.

Dieses Werk sowie alle darin enthaltenen einzelnen Beiträge und Abbildungen
sind urheberrechtlich geschützt. Jede Verwertung, die nicht ausdrücklich vom
Urheberrechtsschutz zugelassen ist, bedarf der vorherigen Zustimmung des Verla-
ges. Das gilt insbesondere für Vervielfältigungen, Bearbeitungen, Übersetzungen,
Mikroverfilmungen, Auswertungen durch Datenbanken und für die Einspeicherung
und Verarbeitung in elektronische Systeme. Alle Rechte, auch die des auszugsweisen
Nachdrucks, der fotomechanischen Wiedergabe (einschließlich Mikrokopie) sowie
der Auswertung durch Datenbanken oder ähnliche Einrichtungen, vorbehalten.

Imprint:

Copyright © 2013 GRIN Verlag GmbH
Druck und Bindung: Books on Demand GmbH, Norderstedt Germany
ISBN: 978-3-656-52422-9

GRIN - Your knowledge has value

Der GRIN Verlag publiziert seit 1998 wissenschaftliche Arbeiten von Studenten, Hochschullehrern und anderen Akademikern als eBook und gedrucktes Buch. Die Verlagswebsite www.grin.com ist die ideale Plattform zur Veröffentlichung von Hausarbeiten, Abschlussarbeiten, wissenschaftlichen Aufsätzen, Dissertationen und Fachbüchern.

Effect of E65 and E70 splice isoforms on electrophysiological properties of Kv10.1

Protocol of laboratory rotation

Maryna Psol, Vincenzo Romaniello

2013

ABSTRACT

Kv10.1, the voltage-gated non-inactivating delayed rectifier potassium channel, is overexpressed in variety of cancer cells and is involved in oncogenesis and tumor progression. Its splice variants E65 and E70, which were discovered in melanoma cell lines, have no conducting abilities but may physically interact with Kv10.1 and thereby regulate its function. Here, we investigated possible influence of E65 and E70 on electrophysiological properties of Kv10.1. Analysis of current-voltage relationships revealed a dose-dependent reduction of Kv10.1 current mediated by both splice variants. The channel demonstrated characteristic feature of activation kinetics, a Cole-Moore shift, irrespective of the isoforms presence. Both E65 and E70 were able to increase the rise time after -60 mV conditioning when expressed at 1:10 ratio with full length channel. Co-expression of E65 or E70 with Kv1.4 did not resulted in considerable changes in channel activity; therefore interactions of splice variants with Kv10.1 are likely to be specific. Downregulation of Kv10.1 activity by E65 and E70 splice variants may modulate tumorigenesis and be associated with less aggressive forms of cancer.

INTRODUCTION

Cancer is one of the major causes of death, amounted to about 12.7 million cases and 7.6 million lethal outcomes worldwide in 2008 (Jemal A., 2011). Changes in the ion channel repertoire are among the hallmarks of malignant transformation (Huber S.M., 2013).

Kv10.1 (Eag1, ether á-go-go) is the first voltage-gated potassium channel whose aberrant expression was shown to be involved in oncogenesis and tumor progression (Pardo L.A., 1999; Hemmelein B., 2006). It was found in diverse cancer cell lines (SHSY-5Y, MCF-7, IGR39 etc) and solid tumors, including sarcoma, colon carcinoma, cervical and gastric cancers (Asher V., 2010), while in healthy individuals Kv10.1 is majorly restricted to brain, placenta and myoblasts of certain stage (Hemmerlein B., 2006). Activation of Kv10.1 facilitates the angiogenesis by stimulating hypoxia inducible factor-1 (HIF-1) and influences adhesion, proliferation and contact inhibition properties by re-organizing cytoskeleton (Asher V., 2010). Moreover, channel activity and surface expression are modulated depending on the stage of the cell cycle: Kv10.1 current undergoes rectification during G_2-M transition and G_0/G_1 arrest (Pardo L.A., 1998; Kohl T., 2011).

Similar to other voltage-gated K+ channels, Kv10.1 consists of 4 α subunits arranged in a circle. Each subunit comprises 6 transmembrane domains (S1-S6) where S4 segment acts as a voltage sensor and S5-S6 domains take part in pore formation. The N- and C-terminal regions are intracellular and may influence kinetic properties and voltage dependence of the channel. The amino terminus includes Per-Arnt-Sim or PAS domain which activates HIF-1, providing selective advantage for tumor cells growth in hypoxic condition. Carboxyl terminus contains a cyclic nucleotide binding domain (cNBD), a calmodulin binding domain (CaM) and a tetramerizing coiled coil domain (TCC), which are necessary for correct channel subunits assembly and perinuclear localization (Ascher V., 2010; Terlau H., 1998).

Activity of Kv10.1 channel can be controlled by various mechanisms: internalization of the channel by endocytosis, control of surface expression, sorting and retention in the endoplasmic reticulum (Kohl T., 2011), and posttranslational modifications in the Golgi complex, e.g. glycosylation (Napp J., 2005). Moreover, alternative splicing was suggested to contribute to the functional properties of Kv10.1 (Ramos Gomes F., 2010) but its' role is not yet completely understood.

Alternative splicing of *KCNH1* gene results in at least two distinct transcripts - E65 and E70 – in addition to the full length forms (Fig. 1). E65 and E70 splice variants were discovered in melanoma cell lines and were named according to their expected weights in kDa. These splice variants are produced by exon skipping: while full length Kv10.1 mRNA is made of 11 exons, E65 and E70 lack 4th to 9th and 4th to 7th exons respectively. Therefore, such isoforms have no transmembrane

3

domains. Also E65 and E70 do not elicit any current when expressed alone in *Xenopus* oocytes (Ramos Gomes F., 2010). Splice isoforms co-express with Kv10.1 in certain cell types: E70 was found in the brain, SH-SY5Y; E65 was detected in DU145, IGR39, and PNT2 cancer cell lines. Both E65 and E70 were shown to colocalize with the full length channel when expressed in the HEK 293 cells. Although, splice variants do not generate functional channels themselves, they may physically interact with Kv10.1 and modulate its' activity.

Fully functional Kv10.1 channel requires N-linked glycosylation in at least two positions: core glycosylation at Asn-388 may be essential for correct folding and stability of the polypeptide, while complex glycosylation at Asn-406 is necessary for proper trafficking and function of the channel. Attachment of core or complex oligosaccharides determines 2 distinct cellular variants of Kv10.1 - of ~110 and 130 kDa respectively (Napp J., 2005). The amount of complex glycosylated Kv10.1 decreases in presence of E70 isoform (Romaniello V., unpublished), and co-expression of E65 causes the reduction of the overall Kv10.1 quantity in the cell according to immunoblotting experiments (Ramos Gomes F., 2010).

Besides, E65 splice variant rectifies Kv10.1 current in the same way as the injections of progesterone or mitosis-promoting factor, and, consequently, E65 may induce maturation (Ramos Gomes F., 2010; Pardo L.A., 1998). Molecular mechanisms of interactions between Kv10.1 and its splice isoforms should be further investigated.

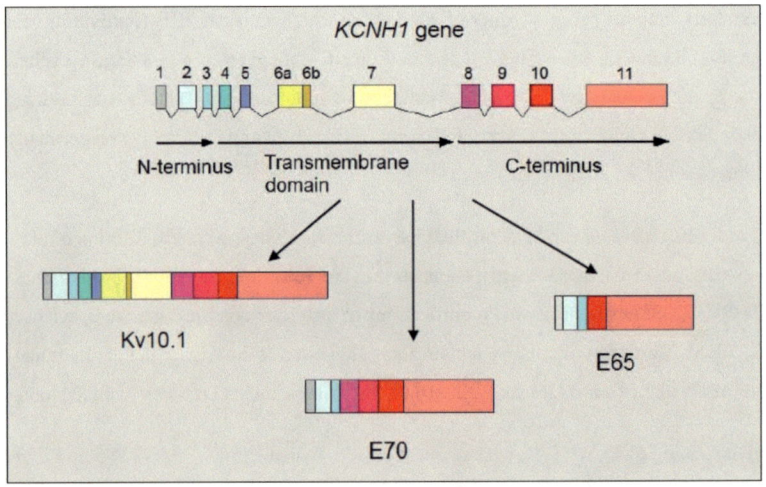

Figure 1. *KCNH1* pre-mRNA can be alternatively spliced into the full length Kv10.1, the E65 isoform, which lacks exons 4-9, or the E70 isoform, which lacks exons 4-7.

Kv10.1 channel has characteristic electrophysiological properties. Its' activity is strongly dependent on holding potential and usually demonstrates rectification at very positive voltages. A specific feature of the channel is so-called Cole-Moore shift, gradual decrease of activation speed after hyperpolarizing prepulse potentials. During long-lasting depolarizing pulses Kv10.1 does not undergo inactivation. (Pardo LA, 2008). In this project we aimed to investigate influence of E65 and E70 splice variants on these electrophysiological characteristics of Kv10.1 channel, dose-dependence and specificity of such effects.

MATERIALS AND METHODS

Synthesis of capped RNA

Capped RNA (cRNA) is an analog of mRNA which is able for translation in eukaryotic cells due to presence of a 7-methyl guanosine cap at the 5'-end. It was synthesized for heterologous expression of Kv10.1 and its' splice isoforms in *Xenopus* oocytes and subsequent electrophysiological recordings.

In order to obtain cRNA we used pSGEM vectors with subcloned template DNA for Kv10.1, E65 or E70 proteins. The quality of the pSGEM-Kv10.1, pSGEM-E65 and pSGEM-E70 plasmids was checked in a control digestion with *XhoI* and *SpeI* enzymes (Fig. 2A).

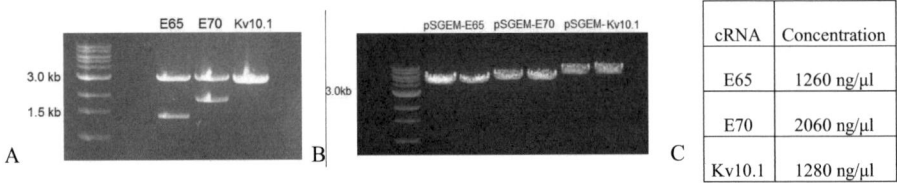

Figure 2. cRNA production. (A) Control digestion and (B) linearization of pSGEM-Kv10.1, pSGEM-E65 and pSGEM-E70 vectors. (C) Obtained concentrations of cRNA.

Sense cRNA was produced with the T7 mMessage mMachine kit. pSGEM constructs (10μg each) were linearized with *SfiI* enzyme in a total volume of 50μl at 50°C overnight. The digestion was terminated with adding 2.5μl 0.5M EDTA, 5μl 3M Na acetate and 100μl ethanol. Samples were cooled at -20^0C for 2 hours, DNA was pelleted by centrifugation (16.400 x g 4^0C 20 min), washed with 70% ethanol, dried and resuspended in 10μl dH$_2$O. Linearization efficiency was examined on 0.8 agarose gel (Fig. 2B).

The concentration of linearized DNA was determined photometrically by measuring absorption of the samples at 260 nm. The ratio A260/A280 was used to estimate DNA purity, and the A260/A280 value about 1.8 indicated lack of protein contamination in the samples.

Transcription reaction reagents were assembled at room temperature in the order and amount indicated in the kit protocol. Then T7 reaction was incubated 2 hours at 37^0C for the maximum yield of cRNA. The template DNA was removed by digestion with 1μl of TURBO DNase at 37°C for 15 min.

RNA was recovered with lithium chloride precipitation. For this we added to each sample 30μl of nuclease-free water and 30μl of LiCl precipitation solution. Afterwards reactions were kept at -20^0C for 30min, centrifuged at 16.400 x g 4^0C for 15min, washed with 70% ethanol. Each sample of obtained cRNA was resuspended in 15μl of DEPC water. Concentrations of cRNA were determined with IMPLEN nanophotometer (Fig. 2C). Then samples were aliquoted and stored at -80°C.

Preparation and injection of *Xenopus laevis* oocytes

Ovarian tissue was surgically removed from mature female *Xenopus laevis* anesthetized on ice. To remove the follicular cell layer the oocytes were treated for 2 hours with 0.2% collagenase solution.

Then they were thoroughly washed in the Barth's medium and selected for injection (Almouzni and Wolffe, 1993).

Oocytes were sorted and injected within 24 h of oocyte preparation. During the fine selection we chose the oocytes of stages V-VI, ~1.2 mm in diameter, two-colored with good separation of dark and light halves, without residues of follicular tissue and visible damage of the cells. The injections were performed with ~50 nl of cRNA solutions with following concentrations:

- 2.5 ng/μl Kv10.1
- 2.5 ng/μl Kv10.1 + 2.5 ng/μl E65 (1:1)
- 2.5 ng/μl Kv10.1 + 25 ng/μl E65 (1:10)
- 2.5 ng/μl Kv10.1 + 2.5 ng/μl E70 (1:1)
- 2.5 ng/μl Kv10.1 + 25 ng/μl E70 (1:10)

- 7.5 ng/μl Kv1.4
- 7.5 ng/μl Kv1.4 + 7.5 ng/μl E65 (1:1)
- 7.5 ng/μl Kv1.4 + 75 ng/μl E65 (1:10)
- 7.5 ng/μl Kv1.4 + 7.5 ng/μl E70 (1:1)
- 7.5 ng/μl Kv1.4 + 75 ng/μl E70 (1:10)

Injected oocytes were incubated at 18°C in ND96 medium with addition of theophylline and gentamicin:

- 96 mM NaCl;
- 2 mM KCl;
- 0.2 mM $CaCl_2$;
- 2 mM $MgCl_2$;
- 5 mM HEPES;
- 0.5 mM theophylline;
- 0.1 mM gentamicin
- pH was adjusted to 7.5 with 5M NaOH.

Two-electrode voltage clamp (TEVC)

The electrophysiological recordings were performed 20-24 h after injection at room temperature, using a Turbo TEC-10CD amplifier (npi Electronics). The intracellular electrodes had resistances of 0.7-1.5 MΩ when filled with 2M KCl.

As an external measuring solution we used NFR, Normal Frog Ringer, which contained:

- 115 mM NaCl
- 2.5 mM KCl

- 1.8 mM CaCl$_2$.
- 10 mM Hepes
- $\pm$3 ml NaOH for 1 l of NFR (to adjust pH to 7.2)

Data were acquired with PULSE (HEKA Electronics) and analyzed with FitMaster (HEKA Electronics) and IgorPro (WaveMetrics) software packages.

In order to characterize current-voltage relationships we applied an I-V protocol, depicted in the figure 3 A. The membrane potential of the cells was held at -80 mV. Discrete depolarizing pulses were given starting from +80 mV up to -60 mV, with decrements of 20 mV. Each pulse was applied for 250 ms, sweep intervals lasted 10 sec, and sample intervals were set up at 100 μs. P/N leak subtraction protocol was used. The mean current from 80-95% of the pulse was plotted against the voltage used to elicit the current response, for measurement and comparison of the current to voltage relationships between different groups of oocytes.

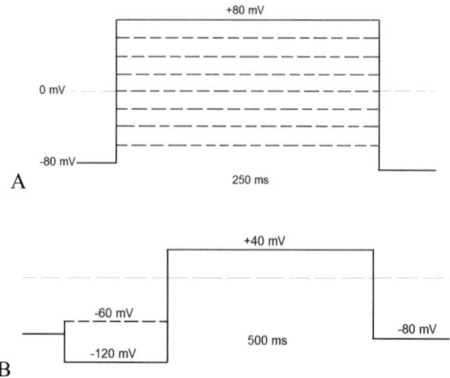

Figure 3. TEVC stimulation protocols: (A) I-V, or current-voltage characteristics protocol and (B) channel activation kinetics protocol.

The channels activation kinetics was measured with protocol described in the figure 3 B. Depolarization pulses of +40 mV were applied for 500 ms after the conditioning pulses (5000ms) of -120 mV and -60 mV. No leak subtraction was made. The rise time of activation was calculated from 20 to 80% of the maximal current.

Statistical analysis

IgorPro and GraphPad Prism software 4.0 were used for the statistical analysis. Differences between groups were evaluated with two-tailed unpaired t-test with Welch's correction. Data are represented as means $\pm$ standard errors of the mean.

RESULTS

Amount of Kv10.1 current decreases in presence of the splice isoforms

In order to evaluate the current-voltage relationships in the oocytes expressing Kv10.1 alone or together with E65 or E70 splice isoforms, we measured current relaxation under a series of depolarization pulses (Fig. 3A). Recordings were performed with 30 – 45 cells in each group.

Co-expression of Kv10.1 with E65 resulted in decrease in total amount of macroscopic current in a dose-dependent manner (Fig. 4). This reduction was statistically significant already when the voltage was clamped at -20 mV. Under the +80 mV pulse the amperage reached the maximal drop on $40 \pm 4.4\%$ ($p < 0.0001$) in oocytes co-injected with Kv10.1 and E65 in 1:1 proportion and on $61.4 \pm 4.5\%$ ($p < 0.0001$) when cRNA ratio was 1:10. The wild-type channels produced relatively linear current-voltage relationship, while addition of E65 led to a slight inward rectification of current at very positive potential.

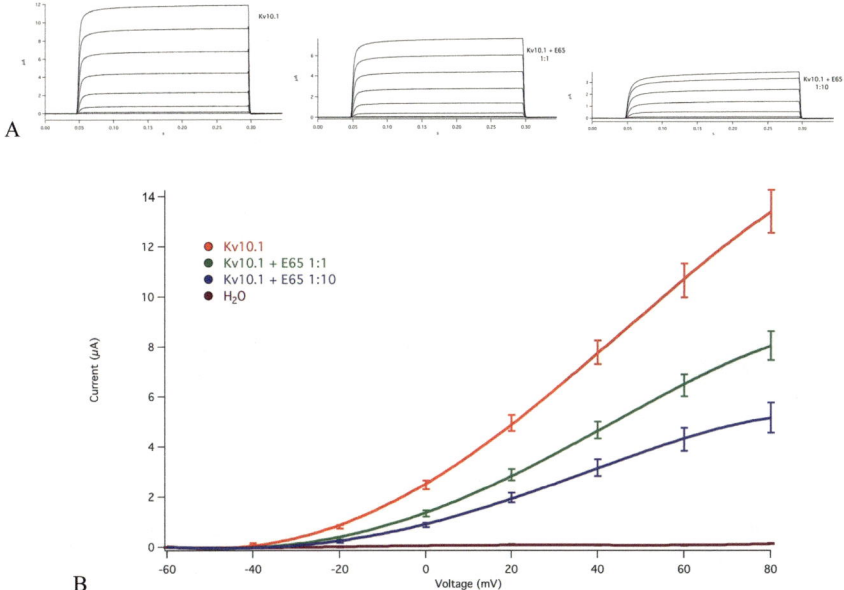

Figure 4. Kv10.1 current decreases in the presence of E65 splice variant

(A) Representative traces of Kv10.1 currents in oocytes injected with channel cRNA and co-injected with Kv10.1 and E65 in 1:1 and 1:10 ratios; (B) Current-voltage relationships; the amount of current at +80 mV significantly decrease when E65 is co-injected with the full length channel cRNA in proportions 1:1 ($40\pm4.4\%$ reduction, $p < 0.0001$) and 1:10 ($61.4\pm4.5\%$ reduction, $p < 0.0001$). Data are represented as means $\pm$ SEM.

Similar electrophysiological profiles were observed when Kv10.1 was co-expressed with E70 (Fig. 5). The reduction of Kv10.1 current amounted to 41.6$\pm$6.7 % (p < 0.0001) and to 66.1$\pm$3.3 % (p < 0.0001) in oocytes with 1:1 and 1:10 ratio of cRNA respectively.

Oocytes, provided with higher quantity of Kv10.1 cRNA (5 ng/µl, 10 ng/µl), demonstrated considerable increase in current magnitude: up to 30 µA at the first day after injection (data is not shown). Therefore, the reduction of current in the cells, which co-expressed full-length channel with E65 or E70, was not caused by the increased cRNA concentration or overloading of protein translation system.

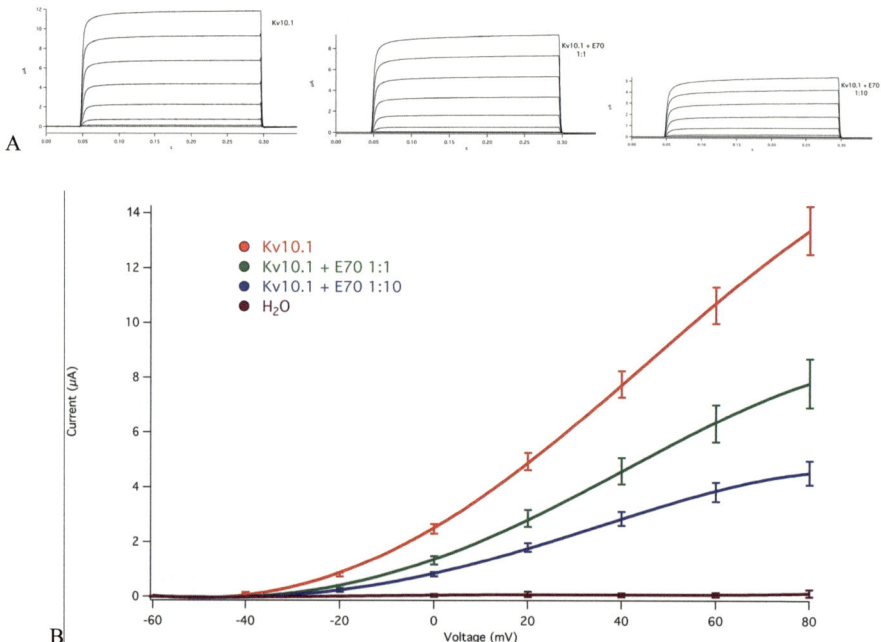

Figure 5. Decrease in Kv10.1 current is observed when E70 is co-expressed in *Xenopus* oocytes (A) Typical current traces elicited with I-V protocol from oocytes, injected with Kv10.1, and co-injected with E70 in 1:1 and 1:10 ratios; (B) Current-voltage relationships; Kv10.1 current magnitude is reduced on 41.6$\pm$6.7 % (p < 0.0001) in presence of E70 in 1:1 ratio and on 66.1$\pm$3.3 % (p < 0.0001) when proportion is 1:10. Data are represented as means $\pm$ SEM.

10

Effect of E65 and E70 on Kv10.1 activation kinetics

One of the most typical electrophysiological properties of the Kv10.1 channel is a progressive deceleration of activation kinetics at hyperpolarizing prepulses which is known as the Cole-Moore shift (Pardo L.A., 2008).

To estimate the influence of splice isoforms on Kv10.1 activation kinetics, we applied depolarization pulse after conditioning at negative holding potentials of -120 mV and -60 mV. Each co-injection group included 19-36 oocytes.

Kv10.1 alone and in combination with E65 and E70 splice isoforms manifested strong dependence of activation kinetics on prepulse potentials (Fig. 6). No considerable difference was observed when splice isoforms were co-expressed with full-length channel in 1:1 ratio. But in 1:10 proportions, both E65 and E70 caused slowdown of the channel activation after -60 mV conditioning ($p < 0.0001$); at -120 mV increase of the rise time did not reach statistical significance.

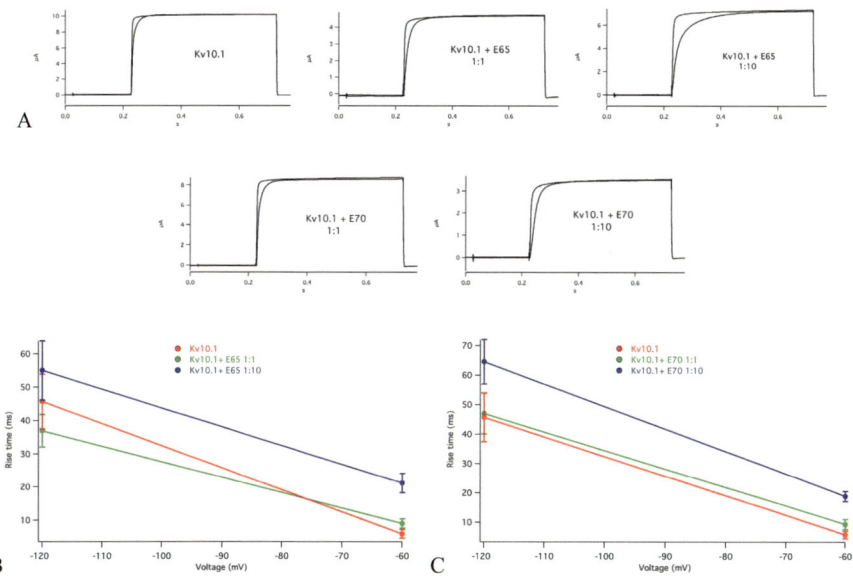

Figure 6. Activation kinetics of Kv10.1 co-injected with E65 or E70 splice isoforms
(A) representative current traces; (B),(C) the rise time of activation from 20-80% of maximal current plotted against the holding voltage; The rise time at -120 mV is not substantially different between groups, while at -60 mV the activation is significantly delayed ($p < 0.0001$) in presence of E65 or E70 in 1:10 ratio. Data are represented as means $\pm$ SEM.

Electrophysiological properties of Kv1.4 in presence of E65 and E70

In order to determine whether interactions between Kv10.1 and its' isoforms are specific we repeated co-expression experiments with Kv1.4 channel. Kv1.4 is a voltage gated potassium channel from Shaker-related subfamily, encoded by *KCNA4* gene (Bett G.C.L., 2004). Unlike Kv10.1, Kv1.4 undergoes rapid N-type inactivation (Fan Z., 2013).

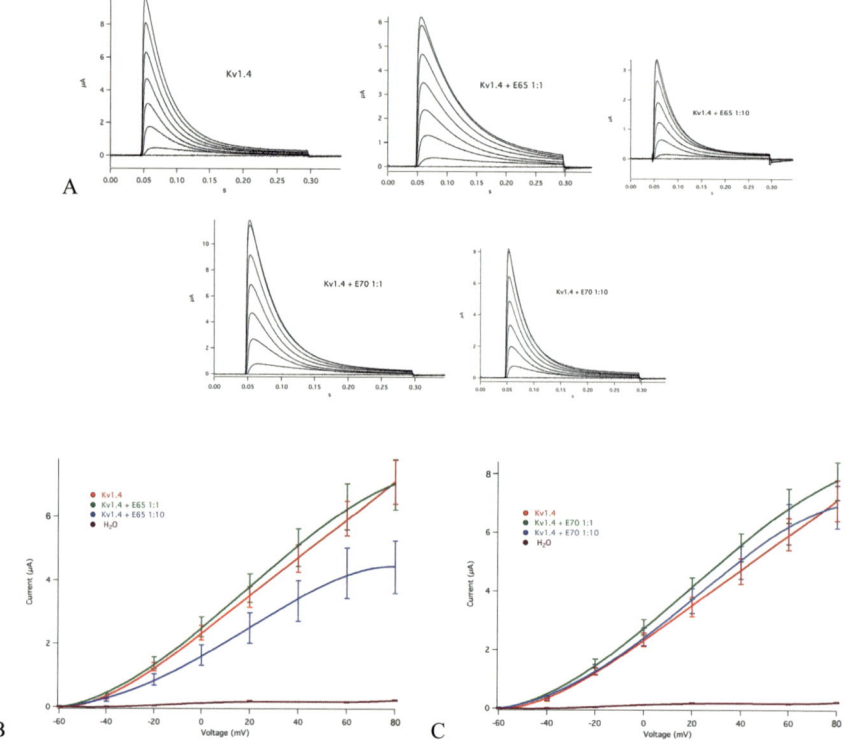

Figure 7. E65 and E70 splice variants do not produce major effect on Kv1.4 currents

(A) Representative traces of currents in oocytes expressing Kv1.4 alone, and with E65 or E70 splice isoforms in ratio 1:1 and 1:10; (B) Comparison of current-voltage relationships in oocytes injected with Kv1.4 (n=36), Kv1.4+E65 in 1:1 (n=29), and 1:10 ratio (n=19), (C) I-V curves from the cells co-injected with Kv1.4+E70 in 1:1 (n=31) and 1:10 proportion (n=24). Kv1.4 current is significantly reduced only at +80 mV in presence of E65 in 1:10 ratio. Co-expression of E65 in 1:1 ratio or E70 does not cause the reduction in amount of Kv1.4 current. Data are represented as mean ± SEM.

E65 and E70 splice isoforms did not produce any considerable effect on Kv1.4 current, except some decrease in amperage at +80 mV in case of Kv1.4 and E65 co-expression in 1:10 ratio (Fig. 7). A slight inward rectification of Kv1.4 current was noticed at very positive voltage in presence of splice isoforms. Although it appeared to be an artifact caused by partial inactivation of the channel by preceding depolarization test pulse, this effect was not observed in the oocytes expressing only Kv1.4. This might indicate slower recovery from inactivation of Kv1.4 in the presence of E65 or E70 isoforms. Overall, the decrease in current magnitude mediated by E65 and E70 splice isoforms is likely to be specific for Kv10.1 channel.

DISCUSSION

Our findings suggest that Kv10.1 current can be strongly downregulated by both E65 and E70 splice variants in a dose-dependent manner. In presence of these isoforms Kv10.1 maintained its characteristic dependence of activation on negative prepulse potentials or so-called Cole-Moore shift. The only considerable change in Kv10.1 activation kinetics was deceleration, observed at holding potential of -60 mV when E65 or E70 were co-expressed with full-length channel in 1:10 ratio. Such effects were not noticed for Kv1.4 channel which indicates specificity of interactions between Kv10.1 and its splice isoforms. E65 mediated reduction of Kv10.1 current is consistent with previously published results (Ramos Gomes F., 2010). Drop of elicited current in the presence of E70 was not reported before, but as opposed to the earlier work of Ramos Gomes F. et al, we used higher ratio of the splice variant to the full length channel and a somewhat different I-V protocol. Here, we for the first time report that the effects of E65 and E70 isoforms depend on their dosage, which may be the mechanism to regulate Kv10.1 channels activity.

Decrease in the amount of Kv10.1 current can be achieved in several ways, e.g. inactivation of channels by biochemical modification, downregulation of surface expression of the channel, interference with processes of synthesis and assembly of the channel, reinforcement of endocytosis etc (Napp J., 2005; Kohl T., 2011). Even though addition of E65 and E70 elicited analogous electrophysiological profiles, the mechanisms, underlying their impacts, may differ. This notion is also consistent with data of molecular biological experiments. Amount of the complex glycosylated Kv10.1 was decreased in immunoblotting under the expression of E70 splice variant (Romaniello V., unpublished). Normal functional state of Kv10.1 channel requires complex glycosylation of asparagine residue 406. N-linked attachment of sugar moieties is also necessary for proper transport and stabilization the protein against degradation (Napp J., 2005). We assume that E70 may inhibit complex glycosylation of Kv10.1 and thereby downregulate the current. Coexpression of E65 with Kv10.1 was shown to be associated with decrease in total quantity of the channel (Ramos Gomes F., 2010). This indicates that E65 may suppress the surface expression of Kv10.1 or stimulate its degradation.

Although two-electrode voltage clamp (TEVC) recordings have proved to be a robust technique to evaluate with high throughput the basic parameters of macroscopic current (voltage dependence, kinetics etc), more elaborated patch clamp recordings are needed to retrieve information on number of active channels, their gating and conductance etc. Other limitations of TEVC method accuracy have also to be considered, such as: artifacts due to variations in cell surface area, series resistance, response time to a voltage command (Dascal N., 2000).

The nature of E65 and E70 impact on Kv10.1 channels may be unraveled with comparison of quantity of active and inactive channels on the cell surface. Ion channel fluctuation analysis could be performed to estimate the number of conducting voltage-dependent ion channels. Opening and closing of ion channels cause nonstationary noise in membrane conductance; the mentioned above technique, based on the noise analysis, allows calculating the number of active channels on membrane, single channel current, and opening probability of the channel (Alvarez O., 2002). General quantity of Kv10.1 channels on the cell surface, regardless of their activity state, could be measured with help of α-bungarotoxin labeling. For this purpose bungarotoxin-binding site has to be genetically introduced into Kv10.1; then the modified channel could be visualized by binding α-bungarotoxin conjugated with fluorophore (Kohl T., 2011; McCann C.M., 2005). Site-directed mutagenesis in C- and N-termini of Kv10.1 may reveal the binding sites for E65 and E70 splice variants.

Kv10.1 changes its conducting properties with the cell-cycle progression. Based on the evidence that maturation and E65 expression evoke similar changes in Kv10.1 current, we speculate that the splice isoforms may participate in modulation of Kv10.1 activity in cell proliferation and oncogenesis.

Taken together, E65 and E70 splice isoforms can cause strong reduction of Kv10.1 current without dramatic changes in such characteristic properties of the channel as Cole-Moore shift. Dose-dependence of the observed effects may underlie possible mechanisms for modulation of Kv10.1 activity and may play role in cancer development mechanisms.

ACKNOWLEDGEMENTS

I would like to thank Prof Dr Luis Pardo for giving me the opportunity to work in his department, and valuable advice on the project. I wish to thank Vincenzo Romaniello for supervising my experiments, and helping with all steps of the laboratory rotation; Ulrike Leipscher for comprehensive explanation of the traces analysis with FitMaster; and Camilo Gomez-Posada for troubleshooting in TEVC method. My thanks go to all members of the group for creating a friendly and inspirational atmosphere. It was my pleasure to work there.

REFERENCES

Almouzni, G. and Wolffe A.P. (1993) Replication-coupled chromatin assembly is required for the repression of basal transcription in vivo. *Genes Dev.* 7: 2033-2047.

Alvarez O., Gonzalez C., Latorre R. (2002) Counting channels: a tutorial guide on ion channel fluctuation analysis. *Advances in Physiology Education* 26 (4): 327-341.

Asher V., Sowter H., Shaw R., Bali A., Khan R. (2010) Eag and HERG potassium channels as novel therapeutic targets in cancer. *World Journal of Surgical Oncology* 8:113

Bett G.C.L., Rasmusson R.L. (2004) Inactivation and recovery in Kv1.4 K^+ channels: lipophilic interactions at the intracellular mouth of the pore. *J Physiol* 556 (1): 109-120.

Dascal N. (2000) Voltage clamp recordings from Xenopus oocytes. *Current Protocols in Neuroscience* 6.12.1 – 6.12.20.

Fan Z., Zhang Z., Fu M., Qi Z., Xiao Z. (2013) Effect of inserting charged peptide at NH_2-terminal on N-type inactivation of Kv1.4 channel. *Biochimica et Biophysica Acta (BBA) - Biomembranes* 1828 (3): 990-996.

Hemmerlein B., Weseloh R.M., de Queiroz F.M., Knötgen H., Sánchez A., Rubio M.E., Martin S., Schliephacke T., Jenke M., Radzun H.J., Stühmer W., Pardo L.A. (2006) Overexpression of Eag1 potassium channels in clinical tumours. *Mol Cancer* 5:41 doi: 10.1186/1476-4598-5-41

Huber SM. (2013) Oncochannels. *Cell Calcium* 53 (4): 241-255.

Jemal A, Bray F, Center M, Ferlay J, Ward E, Forman D (2011) Global cancer statistics. *CA: A Cancer Journal for Clinicians* 61 (2): 69-90.

Kohl T., Lörinczi E., Pardo L.A., Stühmer W. (2011) Rapid internalization of the oncogenic K+ channel Kv10.1. *PLoS ONE* 6(10): e26329. doi:10.1371/journal.pone.0026329

McCann C.M., Bareyre F.M., Lichtmann J.W., Sanes J.R. (2005) Peptide tags for labeling membrane proteins in live cells with multiple fluorophores. *Biotechniques* 38: 945-952.

Napp J., Monje F., Stühmer W., Pardo L.A. (2005) Glycosylation of Eag1 (Kv10.1) potassium channels. *J Biol Chem* 280 (33): 29506-29512.

Pardo L.A., Brüggemann A., Camacho J., Stühmer W. (1998) Cell cycle-related changes in the conducting properties of r-eag K+ channels. *J Cell Biol* 143 (3) 767-775.

Pardo L.A., Camino D., Sánchez A., Alves F., Brüggemann A., Beckh S., Stühmer W. (1999) Oncogenic potential of Eag K+ channels. *EMBO J* 18 (20): 5540-5547.

Pardo L.A., Stühmer W. (2008) Eag1 as a cancer target. *Expert Opin. Ther. Targets* 12 (7):837-843.

Ramos Gomes F. (2010) PhD dissertation, University of Goettingen.

Terlau H., Stühmer W. (1998) Structure and function of voltage-gated ion channels. *Naturwissenschaften* 85: 437–444.